STEEL MILL

STEEL MILL

BY EVE AND
ALBERT STWERTKA

SERIES EDITOR: MELVIN BERGER

INDUSTRY AT WORK

FRANKLIN WATTS · NEW YORK · LONDON · 1978

Photographs courtesy of: Jones and Laughlin Steel Corporation: pp. 2, 5 (top), 6 (bottom), 58 (left); American Iron and Steel Institute: pp. 5 (bottom), 12 (bottom), 15, 22, 26 (left), 28, 33, 41 (right), 50, 57 (right), 58 (right); Bethlehem Steel Corporation: pp. 6 (top), 11, 26 (bottom), 41 (left), 48 (top), 57 (left); United States Steel Corporation: pp. 12 (top), 21, 38 (left), 45; Republic Steel Corporation: pp. 34, 48 (bottom); 61; Interlake, Inc.: pp. 35, 38 (right).

Cover photographs: Republic Steel Corporation (top); Interlake, Inc. (bottom).

Library of Congress Cataloging in Publication Data

Stwertka, Eve.
Steel mill.

(Industry at work)
Bibliography: p.
Includes index.
SUMMARY: Describes how steel is made and turned into such finished products as sheets, pipes, and wires. Discusses some of the research efforts being undertaken by steel plants.
1. Steel—Juvenile literature. 2. Steel-works—Juvenile literature. [1. Steel. 2. Steel industry and trade] I. Stwertka, Albert, joint author. II. Title. III. Series: Industry at work series.
TN732.S8 669'.142 78-2319
ISBN 0-531-02208-0

Printed in the United States of America
6 5 4 3 2 1

CONTENTS

TO MELVIN BERGER, OUR MENTOR

ACKNOWLEDGMENTS

We are extremely grateful to the men and women of the steel industry whose help enabled us to write this book.

Our special gratitude goes to William K. Peck of the United States Steel Corporation's Fairless Works, Pennsylvania, for his hospitality, and for the generous gift of his time and expertise.

We also wish to express our sincere appreciation to Janet Devereux of the New York office of Bethlehem Steel Corporation; Dee Engler of Interlake, Inc.; Charles M. Nekvasil and C. E. Cervenka of Republic Steel Corporation; Jeffrey Given of the American Iron and Steel Institute; Lois England of the Pittsburgh office of United States Steel Corporation; M. D. Post of Bethlehem Steel Corporation; Christine M. Infante of Jones and Laughlin Steel Corporation; James R. Butler of Youngstown Steel; John T. Thompson of United States Steel Corporation, Fairless, Pennsylvania; and R. W. Pasnick of the United Steelworkers of America.

INTRODUCTION

Steel mills are like busy cities. They sprawl across thousands of acres. Enormous buildings stand side by side in orderly patterns. Some of these buildings are as long as twenty football fields. Others are skyscrapers as tall as twenty-five-story towers. Rows of chimney stacks reach to the sky. Between the buildings run miles of roads and railway tracks. One hears the thundering noise of coal and iron ore being unloaded. One hears the roaring of furnaces and the hissing of steam, the hooting of locomotives and the clanking of trains.

Steel mills make their own electricity. They have their own water treatment plants and their own transportation systems. They have a fire department and a hospital. Running a steel plant smoothly around the clock is like maintaining a city of about twenty-five thousand people. There are two hundred steel companies in America. Together, they operate plants in thirty-eight states, coast to coast.

On the job, many workers wear special clothing and protective glasses to guard them from the heat.

Most steel mills stretch along rivers or lakes. Day and night, ships bring in raw materials. They bring iron ore from the Lake Superior region, and also from West Africa, South America, and Labrador. They bring coal from Pennsylvania and West Virginia. At the same time, freight trains bring limestone from Michigan, Pennsylvania, and Ohio. They bring scrap iron from junkyards all over America. Many thousand tons of these materials are heaped on storage piles each day. From here they are transported to the furnaces and turned into steel. A modern plant can make more than 17 tons (15,000 kg) of steel in just one minute.

Being near lakes or rivers is important to steel plants. Lots of water is used in making steel. It is used both for cooling and washing. Before the water is returned to its source, it must be treated, cooled, and cleaned. The same is true of the great amounts of air used in a steel plant. The company must be careful to remove dirt and chemicals coming from its smokestacks.

In the steel yard, employees work near huge machinery and hot metal. They have to watch out for overhead cranes and moving vehicles. This is why steel companies must be alert to safety. On the job, many workers wear steel-capped shoes, hard hats, and protective glasses. Sometimes their hands and faces are masked to guard them from the heat. On every wall, signs remind people to be careful. Dangerous areas are clearly marked with bright paint. Loud horns sound when trucks or trains start to move. Safety experts install safeguards on all machines. Because of these precautions, the steel industry is ranked among the safest of all heavy industries.

After the steel has been made and cooled, it is given various shapes. It can be rolled flat, formed into bars and pipes, or drawn into wire. Then it is delivered to all parts of America and all over the world.

How do the tons of raw materials brought into the plant each day become transformed into those heavy bars, smooth sheets, and shiny wire coils shipped out a few days later?

It takes the teamwork of thousands of men and women who work in the steel plant. Engineers and scientists constantly develop better ways to make and shape steel. Sales managers and production supervisors plan for the kinds and quantities of steel ordered by each customer. Production workers handle the machines and furnaces for this giant task. Maintenance workers keep the machines and furnaces in order. Quality control experts supervise every step. Transportation workers move the raw materials and the finished steel products.

At the head of the steel company are the executives. These are the men and women who carry the greatest responsibility. They make the biggest decisions and therefore earn the highest salaries. Their aim is to keep the company growing. Some executives must know all about places to buy raw materials, others must know about stocks and bonds, and still others must know about the best sites for new steel plants.

Half a million people work at making steel. A large number of these are laborers. They are not very highly

Above: sales manager and department supervisor plan the kinds and quantities of steel to be made. Below: half a million people—women as well as men—work at making steel.

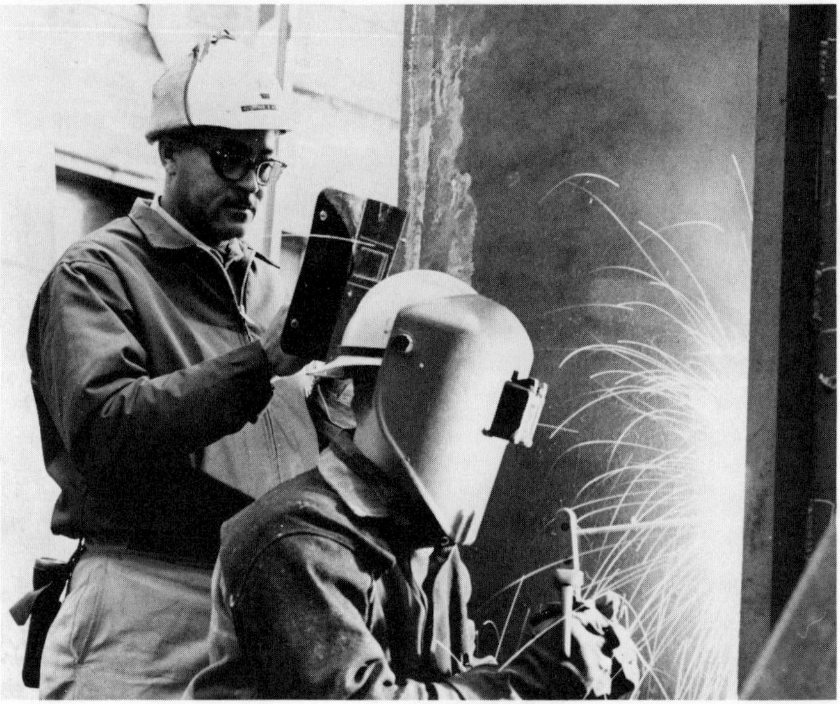

Above: company instructors teach new workers. Here, a welding instructor watches a student doing a practice weld. Right: often, workers start as general helpers. Later, they may advance in their particular job until they are put in charge of the department.

skilled and may not have much schooling. Their work area may be hot and noisy, and their jobs may quickly become routine and boring. They are paid by the hour, and when business is slow, they are laid off. Often, workers start as general helpers. Soon, the company may place them in an apprenticeship program for special training. Company instructors teach them new skills. Later, they may advance in their particular job until they are put in charge of their department. As their skill increases, their wages increase also.

A certain number of people working in steel companies are trained experts with university degrees. They are paid high yearly salaries in return for their knowledge and skill.

About 45,000 of the men and women employed in making steel are members of the United Steelworkers of America. It is the largest union in the American Federation of Labor (AFL-CIO). It is a powerful champion of the workers' health and safety. It has won higher salaries and better working conditions for its members.

One important union issue is to win a secure income for steelworkers who are replaced by machines. In recent years, though, the steel industry has been growing. Therefore, the many marvelous new automatic machines have reduced the total number of workers by only about 8 percent.

No wonder the steel industry continues to grow! Do you realize the importance of steel in your daily life? Cars, bicycles, trains, ships, and planes are all made of steel. Steel is used to reinforce buildings and dams. Bridges are made of steel and so are small familiar objects such as thumbtacks and soda cans. Steel is important in hospitals and offices. Even clothes and food can't be produced without the help of machines made of steel.

No matter where you live, you can hardly pass a day without using many things made of steel. To all of these, the men and women of the steel industry have contributed their skill and their care.

PREPARING RAW MATERIALS

Steel is a mixture of many different ingredients. Each day, these ingredients arrive at the mill by the hundreds and thousands of tons. Large steel companies have their own ore fleets and their own branch railways that bring raw materials to the mill's dockworkers and unloading-platform workers.

At the water's edge, high steel structures called ore bridges jut out over the ships that come to the dock. Ore bridge crane operators ride back and forth along the overhead rails of the ore bridges. They work the cranes that unload the ships. The jaws of the crane buckets can grasp 15 tons (13,000 kg) of raw materials in one bite. The crane operators pile up mountains of raw materials near the dock.

The most important ingredients of steel are iron ore, coal, and limestone. Equally important to the process of making steel are air, water, and energy.

IRON ORE

Today, most of the iron ore mined in the Lake Superior Region of the U.S. and Canada is found embedded in a hard rock called taconite. Before taconite can be used to make iron, the rock is crushed to powder. The iron-bearing particles are removed, and shaped into walnut-sized pellets. In this concentrated state, the taconite pellets are about 65 percent iron.

Even leftover taconite dust is valuable. Workers mix it with coal dust and heat it in special ovens to produce small porous lumps of iron ore called sinter. Every year, about 50 million tons (45 billion kg) of sinter are consumed in blast furnaces.

Mountains of taconite pellets and sinter lie stored near the blast furnaces, together with other materials. They will be loaded or "charged" into the furnaces to make iron.

COAL AND COKE

To be used in the blast furnace, "soft" coal, as it comes from the mines, must first be changed into coke. Coke is pure and hard. Inside the furnace it will not get crushed under the weight of iron and limestone. It will make a very hot fire and will not leave any tarlike, sticky mass.

Coke is made in tall, narrow ovens. The ovens are joined side by side in rows, or "batteries," of up to a hundred or so. They stand in the open air.

Larry cars full of coal ride on rails along the roofs of

An inspector looks over a stockpile of taconite pellets.
The iron content of the pellets is about 65 percent.

Above: after coal has been changed to coke, a door cleaner removes soot and tars from the coke oven doors. Below: coke oven batteries are checked by an inspector, to make sure they are in good order.

the ovens. A larry car operator controls storage bins called hoppers that pour coal into the ovens through holes or ports in the top. The operator allows 16 to 20 tons (14,000 to 18,000 kg) of coal to drop into each oven, and then moves on.

Inside the walls of the coke ovens, gas burns in pipes. This heats the coal to temperatures as high as 2000°F (1093°C). The heat drives off gases, tar, and ammonia, as well as other chemicals. These are useful by-products. They are piped off and sold. They are used for making medicines, plastics, and fertilizers.

A coke oven heater is in charge of keeping the ovens burning at the right temperature. Coking takes about seventeen hours. When coking is finished, a door machine operator removes doors at the front and at the back of the ovens. A pusher operator uses a machine that pushes the red-hot mass of coke out of the oven. The coke falls into a waiting quench car. The quench car operator turns an overhead water spray on the coke to quench or cool it quickly. Then a door cleaner removes baked-on soot and tars from the oven doors.

Coke oven workers probably have the hardest jobs in the steel yard. Their work is hot and sooty. Above all, breathing in smoke and gases can make them ill. After a few years they are given a chance to transfer to other jobs in the mill.

Because smoke from coke ovens may pollute the air all around the plant, inspectors regularly check to see if the ovens are in good order. Steel industry scientists are trying hard to get rid of coke oven smoke. One new method is to collect small dust particles left in the smoke and turn them into pellets for burning.

LIMESTONE

When limestone arrives at the steel mill, the rocks are all shapes and sizes. Workers operating sorting machines crush and screen the stones according to size. The smaller pieces are fed into the blast furnace where iron is made.

The coarser pieces of limestone are crushed and heated. The heat drives carbon dioxide out of the powdered limestone and leaves pure lime. Lime is fed into the furnaces where steel is made.

Both limestone and lime are used as a "flux." This means that the chemical they contain soaks up small pieces of dirt in the molten metal and causes them to float to the top. They form a sort of scum, or "slag," that can be skimmed off.

SCRAP

Iron and steel in every form are too valuable to waste. Lots of iron and steel scrap comes from America's junkyards: auto bodies, cans, machinery, radiators, old ships, and abandoned railroad tracks are all used for scrap. Odd pieces left over after workers shape and trim various steel products are also collected and used again.

Wearing a hard hat and protective shoes, a raw materials handler prepares lime for the steel furnace.

Transportation workers bring scrap to the steel plant by ship and by rail. Here, crane operators manage huge magnets to pick up scrap metal by the ton. They move it from a vast scrap heap to the steelmaking furnaces.

AIR AND OXYGEN

Does it surprise you to know that air is an important raw material for making steel? Even though the air all around us is free, preparing air for the steel plant is expensive. Engineers use complicated filters to clean the air before and after use. Just forcing tons of air to rush into the furnaces is costly. So is heating the air and cooling it again.

Steel is hungry for oxygen. Steel mills use more oxygen than any other industry. At the oxygen plant in the steel yard, workers compress air in large tanks to make it liquid. They separate, or distill, the air into its component parts. Dry air contains nearly 21 percent oxygen and about 78 percent nitrogen. The rest is a mixture of argon and other rare gases. All of these have commercial value, and are collected after distillation and sold.

WATER

For each ton of raw steel they make, workers at the plant use over 37,500 gallons (150,000 liters or 165 tons) of water. Some of this water keeps running through a closed system of cooling pipes. When the water gets too hot, maintenance workers tap it. They run it through tall open structures called cooling towers. When it has cooled, they feed it back into the cooling system. Each time they do this, some water is lost and fresh water must be pumped in from a lake or river.

To keep the water from clogging or corroding pipes, water treatment plant engineers use many methods to filter out dirt and acids. Water used for quenching coke and washing steel must be filtered through a series of pools called clarifiers.

Technicians regularly monitor the quality of the water by chemical analysis and also measure its temperature. Not only that: U.S. government inspectors examine the water steel companies pour back into rivers and lakes. If it contains poisonous chemicals or oil and grease, or if the water is too warm, the company has to pay a large fine.

ENERGY

Steel companies are among the largest industrial users of energy. They get their energy by burning various fuels, such as oil, coal gas, petroleum gas, and natural gas. To make electricity, they burn some of the plant's own gases. These gases are by-products of blast furnaces and coke ovens.

Even so, steel companies must buy more than half of their electric power. Each month, an average-sized steel mill can expect to receive a million-dollar bill from the local electric company.

Hundreds of other products are received by the wharf workers at the plant each day. Some products go directly into the recipe for making steel. Others help in maintaining buildings and machines.

Here is a typical recipe for one ton (900 kg) of steel. Mix the following ingredients:

1 1/5 tons (1000 kg) of iron ore
2/3 ton (600 kg) of coal
1/5 ton (180 kg) of limestone
1/3 ton (300 kg) of scrap steel
8 tons (7200 kg) of air
165 tons (150,000 kg) of water

Heat until molten. Skim off slag. Pour into molds and allow to cool. After cooling, roll to shape.

BLASTING ORE INTO IRON

The first step in making steel is to change the raw iron ore into iron metal. Heat is used to separate the metal from other elements in the ore. This is called smelting. Smelting takes place in giant round towers called blast furnaces. The furnaces are topped by a maze of tubes and pipes.

Blast furnaces can be as tall as twenty-five-story buildings. Usually, two, three, or more blast furnaces stand next to each other. In the blast furnaces iron ore is combined with other raw materials and heated to produce iron.

The raw materials—iron ore, coke, and limestone—are heaped in stockpiles in the yard around the blast furnaces. Nearby is a stockhouse with storage bins. The bins hold the right amounts of material for about one day's operation. A stockhouse operator, standing before a set of dials, controls the mixture of materials that drop from the bins into the skip cars of a conveyor belt. This mixture of materials is called the charge.

Under the supervision of a skip hoist operator, the charge moves up long, sloping ramps to the top of the furnaces. Each skip car dumps or charges its load into the furnace. The mixture of materials in the blast furnace charge seldom varies. Molten iron of the same quality is produced continuously, day after day.

THE BLAST FURNACE

The inside of the blast furnace is lined with special heat-resistant bricks. The temperature can rise as high as 3000°F (1650°C). If you remember that the temperature of the sun is 10,800°F (6000°C) you can understand why special materials are needed to protect the steel shell of the blast furnace.

At these high temperatures the iron becomes liquid. Its glow is so bright that workers need special blue goggles to protect their eyes.

Once the furnace is started, or "blown in," it runs twenty-four hours a day, seven days a week. It runs for about eight years, before being shut down for repairs. If it were allowed to cool off, it would collapse inside. While the furnace runs, one can hear the roaring sound of rushing air, and see showers of sparks and flames.

The way the blast furnace works is very simple. Air is heated in tank-shaped stoves next to the furnace. An operator called the tuyerman (pronounced twee-YER-man) controls huge jet fans that blow hot air into the bottom of the furnace. The air fans the flames to raise the temperature. The tuyerman's main job is to care for the nozzles or

Raw materials lie heaped around the blast furnaces. Next to each furnace are special stoves where air is heated.

Protected from head to foot against heat and sparks, the keeper waits to tap the blast furnace.

tuyers, through which the air enters the furnace. Even a slight blockage of a tuyer must be avoided. The amount of air needed for the furnace is enormous.

It takes about four tons (3600 kg) of air for every ton (900 kg) of iron produced. Can you imagine that the weight of the air used is actually greater than the combined weight of all the other raw materials except water? This huge amount of air is forced into the furnace at speeds of up to 400 miles per hour (640 km/hr). You can see why it is called a "blast" furnace, and why the person responsible for the whole operation of the entire furnace is called the blower.

THE STOVES

The operation of the stoves is entrusted to a stove tender. Stove tenders generally progress to this responsible position from jobs on the furnace crew.

Each stove is so large that it forms a separate building. At the center of each stove, gas is burned in a heavy metal chamber. The chamber is surrounded by thousands of bricks. The bricks are arranged to leave openings or passageways between them. Because of this "checkered" look, the bricks have come to be called the checkerwork. During the heating period, while burning gas heats the bricks, the stove is said to be "on gas."

When the stove is hot enough, the stove tender turns on enormously powerful blowers. Cold air is forced through the checkerwork. It becomes hot as it passes through the openings between the hot bricks. This hot air is then forced into the blast furnace. The stove is now said to be "on blast." After the stove has used up its supply of hot air, the stove tender puts it on gas again.

Several stoves serve one furnace. Some are on gas, while the others are on blast. The stove tender must regulate the stoves so that the air is heated to temperatures as high as 2000°F (1093°C) before being blasted into the furnace.

CLEANING THE AIR

Gas forms in the furnace during blasting. This gas is piped off through complicated tubes at the top of the furnace. It is used as fuel for the stoves that heat the air for blasting. Before this gas goes to the stoves, it must be cleaned. Blast furnace gas always carries dust with it. If the dust were allowed to enter the stove it would dirty the burning chamber. If the dust were allowed to escape, it would pollute the air. Before being burned, then, the gas is led through pipes called "downcomers" into equipment to remove the dust.

Cleaning the gas is the job of dustwashers and their helpers. They control many types of cleaners. One type is called a wet scrubber. It uses sprays of water to wash the dust out of the gas. Another cleaning method is the baghouse. Gas is sucked into the baghouse in much the same way as air is sucked into an ordinary vacuum cleaner. Inside, the gas passes through dozens of nylon filter bags that trap the dust.

In an average month, more than fifty railroad cars of dust are collected by a steel plant.

CASTING THE FURNACE

When iron ore, limestone, and coke are charged into the blast furnace, they do not immediately fall through the cylinder inside. Blasts of hot furnace gases rush up from below. They hold as much as 275 tons (250,000 kg) of materials suspended in the upper part of the furnace. Slowly the charge works its way down. During this slow drop through the hot gases, the iron is freed from the ore.

As the charge drifts down, different chemical reactions take place. In the top half of the furnace, a great deal of oxygen is removed from the iron ore. About halfway down, the limestone begins to react with the impurities in the ore and coke. It forms a material that looks very much like lava from a volcano. It is called slag.

At the bottom of the furnace there is a large brick-lined storage area called the hearth. It is the hottest part of the furnace. The temperature is well over 3000°F (1650°C). It is hot enough to melt both iron and slag.

The molten iron and slag drip through the burning coke and start filling the hearth. Slag is lighter, and it floats on the iron. After some time, a pool of molten iron is formed. It is about four or five feet (1.2 or 1.5 m) deep. A layer of slag floats on top of it. The furnace keeper and a helper take several samples of the molten iron for temperature checks and chemical analysis.

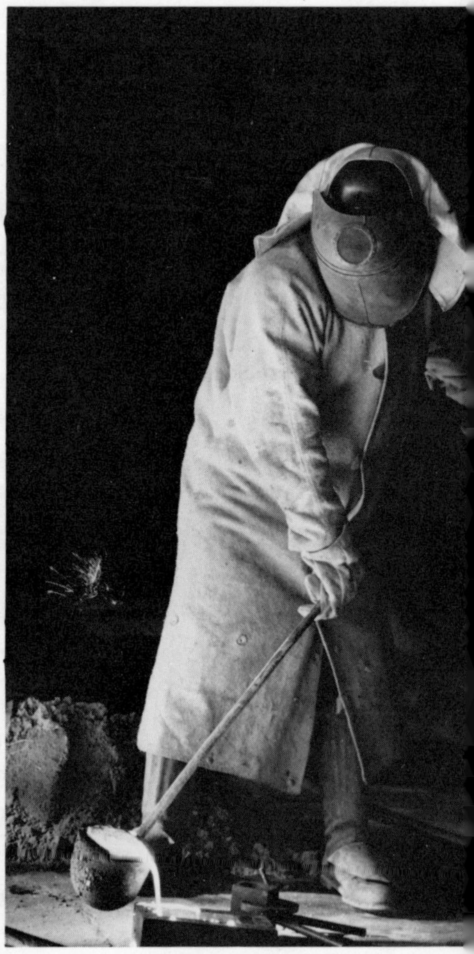

Left: tapping the furnace is a spectacular event. The growing iron with its sparks and flames looks like an erupting volcano. Right: iron samples are analyzed in the chemical laboratory. A well-protected helper dips out a cupful of molten iron.

When the blower and helpers decide that the iron in the hearth is ready, they call for it to be tapped or cast out of the furnace. Usually, a furnace is cast every six hours. The largest furnaces can produce as much as 1000 tons (909,000 kg) of iron per cast.

The keeper and the keeper's helper do the actual tapping. They are protected from head to foot by shiny, aluminized protective suits. To start the cast they drill through a taphole in the hearth, above the molten iron. Slag rushes out into a container called a slag ladle. The ladle looks like a big kettle mounted on a railroad car. Slag contains most of the impurities of iron ore. It is used for making cement or as a material in road building.

After the slag has been removed, the keeper opens a lower taphole. This lets the molten iron pour out. Tapping is a spectacular event. The glowing iron with its sparks and flames looks like an erupting volcano. The iron flows through a channel into insulated "hot-metal" railroad cars. They are shaped like drums and are lined with heat-resistant bricks. They work like gigantic thermos bottles. They keep the molten iron from cooling and becoming solid. Each hot-metal transfer car can hold about 160 tons (145,000 kg) of iron. A locomotive conductor drives the cars by rail to the steel furnace shed. The freshly made iron will now be changed into steel.

PIG IRON

Not all of the iron from the blast furnace is used for steel. Some of it is used for cast iron. It is made into such

Usually, a blast furnace is cast every six hours.

products as radiators, potbellied stoves, bathtubs, and auto engines.

Some of it is sold directly to other foundries in a form known as "pig iron." It is called pig iron because steel mills, in the last century, would pour the molten iron into short channels dug in the ground. They used one main channel with other channels on either side of it to allow the iron to cool. It reminded people of newborn pigs suckling. The central channel was known as the "sow." The side molds were called "pigs."

REFINING IRON INTO STEEL

When pig iron comes from the blast furnaces, it still contains many unwanted substances. These impurities weaken the metal. They make it brittle. To make good steel. the impurities must be removed. This is called refining.

Steel is basically a mixture of iron and a small amount of carbon. Actually, though, there are many different kinds of steel. Steels for surgical tools are different from steels for auto parts. Steels for space satellites are different from steels for water pipes. Some kinds of steel need to be very hard, others, very lightweight; some need to be bendable, others rigid. To get the quality just right, steel is alloyed, or mixed, with small amounts of other metal ores. Alloy steels are often called the royal family of steel.

Some alloy ores are expensive and hard to find. Many have to be shipped from faraway places. They come from Africa, China, and India. The company's chemists use alloys such as manganese (for extra toughness), chromium (for rust resistance), tungsten (for heat resistance), and molybdenum (for corrosion resistance).

To refine iron into steel, several types of furnace are used. The three most important furnaces are: the basic oxygen furnace (BOF), the open-hearth furnace, and the electric arc furnace. All three types of furnace clean the steel by burning out impurities.

More than half of all the steel in this country is made in BOFs. They work quickly and efficiently. The BOF can make 300 tons (272,000 kg) of steel in 45 minutes. The older open-hearth method takes five to eight hours to process the same amount.

The BOF is a huge steel container lined with heat-resistant material. It is shaped somewhat like a pear, open at the narrow end. Workers charge raw materials and oxygen through this opening. The furnace is supported in such a way that it can be easily tilted and turned.

Most steel mills build BOFs in pairs. One furnace can be making steel while the other is being charged with raw materials.

Customers' orders for a certain kind of steel are sent from the sales department to the supervisor of production management. The supervisor sets up a production schedule. He or she also sends an exact recipe of quantities and ingredients to the manager of the furnace division. From here, the order goes to the melter, who directs the making of each batch or "heat" of steel.

The melter works at some distance from the furnace, in an air-conditioned, windowed control room called the pulpit. Furnace operators and several helpers work under the melter's direction. A metals expert, or metallurgical observer, is always present to give advice.

To start, the furnace operator's helper uses levers to tilt the furnace toward the charging platform. A scrap crane operator charges the furnace with scrap steel. Next, a charging crane operator maneuvers a ladle filled with molten iron from the blast furnace. Guided by hand signals from a helper on the ground, the crane operator adds the iron to the BOF. This molten iron makes up 60 to 80 percent of the charge.

When the furnace has been returned to an upright position, the furnace operator lowers a long thick pipe into it. This pipe is called an oxygen lance. It is about 60 feet (18 m) long, and 10 inches (25 cm) in diameter. It blows pure oxygen into the furnace at speeds greater than the speed of sound.

The furnace operator positions the water-cooled tip of the lance so that it is near the surface of the molten charge. When the operator switches on the oxygen, it rushes into the furnace at enormous speed and pressure. As soon as the oxygen comes near the molten metal, it starts to burn, giving off tremendous heat.

By operating the control panel, the furnace operator releases a measured amount of lime into the furnace. Lime helps clean the metal by forming a slag. The slag floats to the top and is later removed. The furnace operator and assistants must regulate the oxygen flow with care. If the mixture overheats, boiling steel and slag splash over the top.

The oxygen "blow," which lasts about twenty min-

The melter supervises the basic oxygen furnace
from an air-conditioned control room.

Hand signals from the helper guide the crane operator, who adds molten iron to the basic oxygen furnace.

The oxygen furnace can be tilted. An assistant takes samples of the "melt" for temperature check and carbon analysis.

utes, refines the molten iron. At the end of the blow, an assistant takes samples of the melt for a temperature check and carbon analysis.

When tests show that the mixture is ready, the helper tilts the furnace toward the "teeming" platform. Molten steel rushes or teems through a taphole halfway up the furnace. It teems into a ladle mounted on a railcar. A mixer operator then feeds the needed amounts of alloying materials into the ladle through a chute. Less than fifty minutes have passed from the charging of the furnace to the pouring of the finished steel.

OPEN-HEARTH FURNACE

The open-hearth furnace is an older and slower method for making steel. It is still used, though, because it can make large batches of steel—up to 600 tons (545,400 kg) at a time. The open-hearth furnace gets its name from the vast fireplace area, called a hearth, where the steel is made. Each hearth looks like a shallow box as big as a basketball court, but only two or three feet (.6 or .9 m) high. Along the front of the hearth, doors can be opened for charging and observation. At the back of the hearth, the taphole overhangs a sharp drop to a lower level.

Usually, two or three open-hearth furnaces are built side by side, along the center of a mammoth shed, several city blocks long.

Stoves at both ends of each hearth create sweeping currents of hot air. The hot air mixes with burning fuels, creating sheets of flame that cover the surface. Many open-hearth furnaces now use oxygen to raise the temperature even higher and to speed up melting.

A charging machine operator begins the refining process by charging raw materials into the furnace. The charging machine runs on tracks along the row of furnaces. An operator rides in an enclosed cabin at the center of the machine. The operator controls a long steel arm called a "peel." It is used to pick up steel boxes filled with measured amounts of scrap and lime. The machine empties these through doors into the furnace.

Now the first helper controls the flow of hot air and fuel from the stoves. Flames shoot across the solid raw materials in the hearth. The fire reaches a temperature of 3000°F (1650°C). It takes about two hours to heat the charge and to melt it.

When the observer and first helper determine that the melt is hot enough, the first helper signals a crane operator. The crane moves along rails fixed in the ceiling. The operator is ready with a ladle full of molten iron from the blast furnace. It has been tipped into the ladle from the hot-metal transfer car. The ladle of glowing iron travels slowly across the shed. Then the operator tilts the contents into the furnace.

The molten mixture may continue to boil for as long as ten hours. During this time, the helpers and the observer inspect the mixture through an open door. Inside, the furnace is white-hot, with a glare as bright as the sun. It can only be looked at through very dark, heat-resistant goggles.

From time to time, a helper, muffled in aluminized

Left: the first helper signals, while a crane operator tilts the ladle, and pours molten iron into the open-hearth furnace. Right: the steel pourer opens a valve at the bottom of a ladle full of hot steel. Metal begins to "teem" from the ladle.

protective clothing, reaches a long-handled dipper into the boiling melt, and pours out small samples of steel. The samples are rushed to the control laboratory. A technician analyzes them and returns the results to the furnace department.

When all experts agree that the melt is ready, the first helper gives the signal for tapping. Warning alarms hoot through the building. Workers nearby hurry for cover.

The taphole at the back of the furnace is about six feet (1.8 m) long, and nine inches (23 cm) in diameter. It has been plugged with hardened clay. The second helper uses explosives to blast out the taphole. A loud bang rocks the building. A torrent of white-hot steel rushes out, among showers of flying sparks. The steel pours into a ladle waiting on the lower level behind the furnace.

ELECTRIC ARC FURNACE

The electric arc furnace is the only steel-making method that does not use molten pig iron from the blast furnace. It makes steel directly from enriched iron ore and carefully selected scrap steel.

The electric arc furnace is a steel cylinder with a domed roof. It can be tilted for tapping. The molten steel pours through a spout on one side. This makes the furnace look like a giant's tea kettle.

Large carbon rods, called electrodes, pass through the roof into the body of the furnace. When electric current flows through the electrodes, lightninglike arcs jump between them and through the scrap metal in the furnace.

The heat of the sparks melts the metal. Even the electrodes turn white hot.

The furnace operator in the control room moves a crane to lift the furnace roof, electrodes and all. The roof swings to one side, and the furnace is charged with scrap steel and lime. The operator replaces the roof, and uses controls to lower the electrodes to within a few inches or centimeters of the steel. Then the operator turns on a power supply as high as 120,000 volts. A roar fills the furnace room.

The operator can use the electric current to regulate the furnace temperature very precisely. Therefore, the electric arc furnace is used for making high quality special steels that need exact temperature control.

INGOTS

The end product of all steel-making furnaces is hot, liquid steel. Before being sent to the finishing mill, the steel must become solid. Hot steel is molded into so-called ingots. These are metal blocks, about 8 feet (2.4 m) long and 3 feet (1 m) wide. The largest ingots weigh up to 600,000 lbs (272,000 kg).

Ingot molds stand lined up on railcars along the pouring aisles of each furnace. A ladle crane operator positions the ladle of hot steel over the molds. The steel pourer opens a valve at the bottom of the ladle. Red-hot steel teems into the molds.

When the metal has cooled in the shape of the mold, it is called an ingot. An ingot stripper uses the grippers of

Left: ingot molds stand lined up on railcars along the pouring aisle. Steel pourer and observer watch as hot steel teems into the molds. Right: a heater watches while the crane operator uses giant tongs to grip the red-hot ingot and lower it into the soaking pit.

an overhead crane to lift off the molds. Rows of finished ingots are left standing on the railcars, ready for transportation to the "soaking pit."

The outside of an ingot cools faster than the inside. This means that the outside can be solid while the inside is still liquid. To even out the temperature, the ingot is "soaked"—not in water, but in heat. This is done in a large, sunken heating tank, or pit. At the pit a crane operator grips each ingot with giant tongs. The tongs lift the ingot and lower it into the soaking area. A heater and helper adjust controls to maintain the correct temperature.

Ingots are usually reheated or soaked until they are about 2200°F (1200°C) all the way through. At this temperature they are solid but soft enough to be shaped. When they are ready, the heater tells the crane operator to lift them from the soaking pit onto a special ingot buggy. A locomotive engineer drives the ingot buggies, one at a time, to the finishing mill.

SHAPING STEEL

At the finishing mills, steel is pressed into many different shapes. Mammoth machines in sheds covering hundreds of acres roll, pound, squeeze, or draw out the metal. Other machines coat some of the metal with tin, zinc, or paint.

Finishing mill machines are marvels of modern technology. For the most part, they work automatically. Engineers have preset or "programmed" them to use their amazing power fast and efficiently.

Each particular machine is supervised by a highly trained operator at a control panel. In many cases, the operator guides the machine by inserting a punched program card into a slot on the control panel. At the push of a button, the machine "reads" the instructions on the card and begins to do the required job. The operator must constantly watch dials and gauges, though, and be alert for any trouble or signs of breakdown in the machine.

Ingots coming from the furnaces can weigh as much

as 600,000 lbs (272,000 kg). They are too bulky to be easily handled. Before they can be "finished" in certain shapes, they must be reduced to more manageable forms. This is called roughing.

At the roughing mill, machines roll and squeeze ingots hot from the soaking pit to make them longer and narrower. Roughing mill operators use their machines to turn out various steel forms of different thicknesses. The most common forms are blooms, billets, bars, and slabs.

Blooms are bulky beams, either square or rectangular in cross section. Billets are narrower in cross section, measuring about six inches (15 cm) square. Billets will be rolled once more and turned into round or square bars. Bars may later be hollowed out to make pipes, or rolled into rods and drawn into wire. Slabs are thick and wide. They will be rolled flat to make plates, sheets, and strips.

The person in charge of the rolling mill is called the roller. The supervisor of the production control department tells the roller what size and quantity of blooms, billets, and slabs have to be made each day. The roller selects the right punch card for the order and puts it in a slot in the control panel. Switches and levers start the machine rolling.

As each red-hot ingot comes from the soaking pit to the conveyor, the machine pulls it between two grooved rolls that turn in opposite directions. The amount of pressure and heat between the rolls is measured by sensitized equipment. The operator controls the pressure and the speed of the rolling. Often, adjustments must be made. Under the operator's guidance the ingot moves forward

A hot slab enters the first roughing stand.
It will be flattened into a long steel sheet.

under several sets of rolls, or roughing stands. Stand after stand presses it longer and thinner. Quickly it is pressed into a long bloom, or billet, or slab. It continues on the conveyor toward the station of a shear operator. The shear operator controls a powerful instrument that cuts the metal into accurate lengths.

CONTINUOUS STRAND CASTING

Some modern plants use a more efficient way to process molten steel. It avoids making ingots and soaking them in heat. Instead, it molds, or casts, a continuous strand of metal.

For continuous strand casting, ladle crane operators from several steel furnaces pour load after load of molten steel into an enormous reservoir called a tundish. From the tundish the liquid steel flows in a steady stream through a water-cooled mold. As the metal flows, sprays of water quickly cool it on the outside. The steel forms a skin and takes on the shape of the mold. The strand of hardening steel gradually bends until it lies flat along the bed of the casting machine. From here, it moves right on to the continuous strip mill where roller machines flatten it into sheets and strips of any desired thickness. A continuous strand caster can convert 300 tons (272,000 kg) of molten steel into solid slabs within forty-five minutes.

SHEET AND STRIP MILL

Sheets and strips of steel are among the most important products of a steel plant. Other factories will use them to make many things, such as auto fenders, kitchen cabinets, and washing machines.

A continuous strip mill with all its subdivisions can take on a 12-ton (10,900-kg) steel slab, 7 inches (17½ cm) thick, and press it into a thin sheet within a few minutes. It can turn out coils of hot rolled sheets at the rate of 3500 feet (1060 m) per minute.

Guided by only a few supervisors stationed at control pulpits, the continuous strip mill coats the sheets with tin or zinc, dips them in paint, dries the coatings, bakes them, and cools them. Finally, it cuts the sheets and winds them on reels to be shipped out of the plant. Often, the roller uses computers and TV cameras to monitor the progress of the sheets through the mill.

A typical strip mill is a shed about a mile and a half (2.4 km) long. Through the center, from one end to the other, runs a conveyor. Each red-hot slab first passes through several sets of roughing stands. With a thunderous noise, the slab resists, at first, and then plunges between the rolls. Each stand of rolls presses harder and makes the slab longer.

The roller uses a computer to adjust the gauge or pressure between rolls. A speed operator controls the speed at which the sheets pass along the conveyor. The roller and speed operator are highly experienced workers. If their adjustments are not perfect, sheets can buckle or tear, and damage the machine.

By the time the glowing hot slab reaches the middle of the shed it has become about 50 feet (15 m) long. Even from a distance one can feel its breathtaking heat. As it becomes thinner, it travels through each rolling stand

Above: guided by an operator, steel sheets pass through the galvanizing line to be coated with tin or zinc. Below: in the shipping room, workers prepare coils of sheet metal for transportation out of the mill.

faster and faster. Now it rushes through a series of pickling tanks full of acid. They remove the rust and scale formed by rolling. At a speed of 3500 feet (1000 m) per minute it goes plunging under a line of water showers that wash it and quickly bring down the temperature.

At the end of the line, the thin ribbon of steel sheeting, now 2300 feet (700 m) long, disappears into a coiling machine. Only seconds later, the machine deposits a huge, shiny coil of steel sheet onto a conveyor. While the finished coil travels slowly along the ground, a coil bander puts a thin steel band around it. Another worker paints a number on each coil with a long brush. The coils travel onto the flatbed of a railroad car, and are shipped to the next building.

Other finishing procedures are in store for the coiled metal sheets. They may be rolled unheated (cold rolled), which makes them even thinner, smoother, and stronger. Furnace workers may reheat them to make them soft and pliable. This is called annealing. Other workers may harden the surface, in a process called tempering.

Steel that will be used to make "tin" cans or roofing material must be "galvanized." This means that it is coated with tin or zinc. The galvanizing machine operator dips the sheets into pools of hot liquid zinc or tin. The machine coats, dries, and coils the sheets.

To give the metal a pleasant appearance, customers often order a coating of paint or vinyl. For this purpose, the sheets of metal pass through a high-speed continuous paint line. Then they pass through air dryers and ovens where the coating is baked on.

A transportation worker uses powerful lifting equipment to move wire coils into the shipping room.

PIPES AND TUBES

Water mains, oil pipes, and gas lines need tubes that are huge in diameter. Tubes that are large enough for these uses are made by bending the sides of steel plates upward and welding the seams together.

Extra strong tubing, though, is made without seams. Seamless tubes and pipes are made at the piercing mill, by hollowing out the inside of a steel rod. Actually, the piercer machine operator does not bore a hole through the center of the rod. Instead, the operator uses the machine to apply heat and rotating pressure. This causes the inner core of the rod to become hollow. It is a little like rolling a string bean back and forth between your fingers until the whole length of the inside softens and opens to form a tube. The piercer operator then inserts an instrument through the steel pipe opening to shape it evenly.

RODS AND WIRE

Rods and wire are made from narrow billets. Under the supervision of a rougher and a roller, the billets are pressed into long, thin rods. Workers clean the rods in a solution of hot acid. They rinse the rods in water and coat them with a baked-on layer of lime.

Conveyors bring coils of treated rods to the wire drawer. The drawer operates equipment that draws the rods through narrow openings, or dies. Tremendous power is needed to draw the wire through these narrow openings. The power is supplied by a thick steel spool called a draw block. The draw block turns steadily. It pulls the wire through the die and winds it around itself into a long coil. The wire goes from one drawing frame to the next. Each frame draws the wire thinner.

Wire has a thousand uses—from bridge cables to guitar strings. Some wire is galvanized for protection or coated with paint for decoration.

Transportation workers use powerful lifting equipment to move the coils of finished wire to the shipping room. From there, trucks and railroads deliver it to all parts of the world.

QUALITY CONTROL AND BASIC RESEARCH

Scientists and engineers of all kinds work in a steel mill. Some are experts in quality control. They sample and test everything from raw materials to finished products. They cooperate with supervisors and workers in all departments—at the ore docks, in the furnace rooms, and at the rolling mills.

Other scientists and engineers devote their time to basic research. They do their work in company laboratories, computer centers, and libraries.

Quality control and basic research are closely related. The aim of both is to make new and better kinds of steel.

QUALITY CONTROL

To make each batch of steel exactly right, chemists test the separate ingredients before letting them be mixed together. Because most of the ingredients are used in enormous quantities, cleaning and sorting them is a difficult operation. Any mistake could mean a ruined batch of steel—a costly and unpleasant surprise.

Among the experts most vital to the steel industry are those who have studied the chemistry and physics of metals. Metallurgists, as they are called, usually have university training. They know both the theory and the practice of metallic reactions. They form a team of specialists, each with a particular area of metallurgy he or she has studied in depth.

Each heat of steel made in the furnaces must be tested several times before casting. In the furnace room, a helper places the test sample into the opening of a tube near the control desk, and closes the cover. The tube connects the furnace room with the quality control department. An air suction device whisks the container that holds the sample through the tube. It lands right at the desk of a metallurgical technician.

Using a modern instrument called a spectroscope, the technician obtains a fast and correct analysis of the sample. The technician adjusts the spectroscope to show the elements in the sample. The scope throws patterns on a screen. These patterns are automatically compared, inside the spectroscope, with another specimen of the desired composition. The technician quickly reads the scope. Then he or she communicates the results by teletype machine to the furnace operator in the control booth.

Quality control engineers take special care in testing finished steel mill products. After all, hundreds of lives depend on the strength of an airplane wing or of a bridge girder.

One recently developed testing method makes use of an ultrasonic device. By means of ultrasonics an operator can detect air bubbles or pieces of foreign matter trapped inside a piece of steel. The operator sets the ultrasonic equipment to send out a pulse. The pulse is a kind of sound that cannot be heard by the human ear. This pulse goes through the material to be tested. It is reflected from the opposite side of the equipment and the reflection appears on a video screen. In places where the steel is thin, the pulse is reflected quickly. Where the steel is thick, the pulse takes longer to penetrate and return. The operator explores the thickness of the steel by measuring the length of time between each pulse and its reflection.

A skilled tester can pinpoint uneven areas in a steel product. Ultrasonic testing makes it possible to find faults as far as 30 feet (9 m) along the length of a steel bar.

Radiography is another modern testing method used by quality experts to test finished metal products. It works very much like the X rays doctors use on the human body. The radiographer uses radiation to make a shadowgraph of the inside of the piece being tested. The radiation instrument casts the shadowgraph on a screen. The operator is trained to see if the test piece has any faults.

Sometimes, metal objects may develop a crack. To remove or repair the damaged part, it is necessary to find out the size and shape of the crack. To do so, a technician treats the crack with glowing dyes. The dyes are absorbed by the tiniest opening. In this way, even a hairline crack becomes visible to the inspector.

MAINTENANCE

Steel-making machines are often damaged by heat and heavy pressure. To make high quality steel, the machines must be kept in good shape. Workers are needed to rebuild, repair, and install all kinds of equipment.

Shop electricians repair worn or damaged electrical tools. Welders are needed for many maintenance jobs. They use different kinds of torches and welding electrodes to join metal pieces together. Bricklayers build and repair the inner linings of furnaces, using special heat-resistant refractory bricks. Carpenters at steel plants construct frames and scaffolds for other workers to walk on. Roll turners repair rolling mill equipment. They operate roll lathes and grinders to smooth and shape the metal rolls exactly to the right dimensions. Pipe fitters install and maintain all pipelines, fittings, and fixtures. Toolmakers replace broken tools.

Each department of the plant has its own maintenance workers who have learned their skills on the job. The millwright maintains mechanical equipment such as rolling mills or cranes in a particular department. The oiler lubricates all mechanical and electrical machinery to make it run smoothly. Trucks, tractors, locomotives, weighing scales, and railroad cars are all maintained by mobile equipment mechanics.

Left: maintenance is part of quality control. Here, a welder repairs the end of a propane tank. Right: a female "set-up man" prepares a production machine for operation.

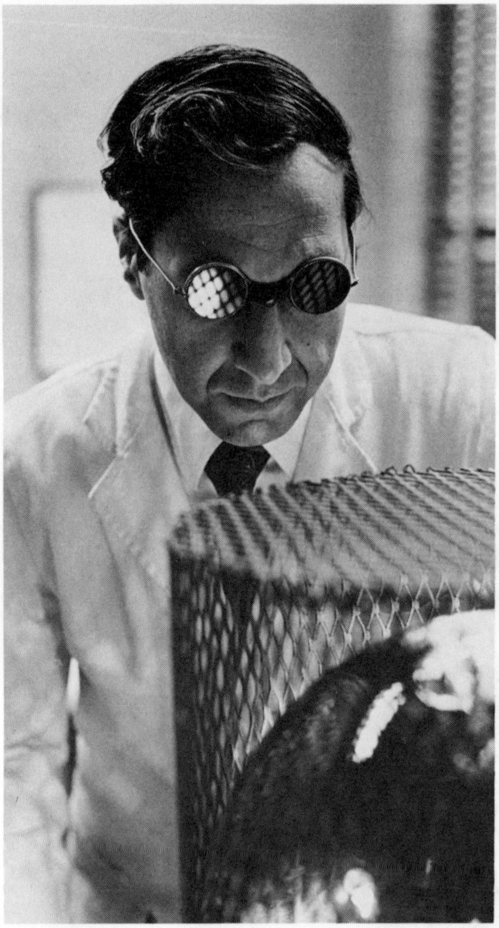

Left: metallurgists study the chemistry and physics of metals. Here, a scientist prepares samples for analysis. Right: to examine the microstructure of iron and steel, a scientist uses an electron microscope.

BASIC RESEARCH

Scientists doing basic steel research need more than test tubes and microscopes for their work. They often need huge equipment such as stoves and furnaces. This is why the laboratories of steel companies are likely to be actual pilot plants—small plants that lead the way in trying out new methods to manufacture steel.

On the other hand, not all steel researchers work with huge machines and mountains of materials. Other researchers concentrate on the secret structure of metal in its tiniest particles. To study the atomic structure of iron and steel, scientists use an electron microscope. With this complicated instrument they can look at tiny bits of metal, enlarged approximately 10,000 times.

Many research instruments used in the steel industry were invented by the company's own engineers to solve a specific problem. Inventing and building new instruments and machines is another part of basic research.

RELATED RESEARCH

Some research at steel plants is only indirectly related to metal and metal products. The plant's chemical engineering staff may be working on other materials, such as plastics. Plastics are used as coatings and wrappings for many steel mill products.

Today, steel plants are also devoting much of their research effort to the control of water pollution and air pollution. Physicists, chemists, biologists, and engineers join forces to give the cities and states where steel plants operate a clean environment.

Company engineers are busy developing more efficient air scrubbers and water-filtering systems. Laboratory technicians check air and water samples to keep track of any pollutants that might be released from the steel mill.

INFORMATION SERVICES

Both quality control experts and basic researchers are helped by the men and women who collect, process, and store information.

Facts and figures, often called "data," are collected on every batch of steel made. Over the years, this information becomes a valuable help in the improvement of steel. Computer operators store the data in large computers. Data analysts and other mathematical experts, such as statisticians, study the data to make new discoveries about steel.

The study of data enables scientists and engineers to predict things about steel products before they are made. This is a less wasteful way to make new types of steel than preparing one batch after another to see how it will turn out.

In the control room of a large computer, one sees spinning disks and hears the hum of electronic equipment. There are not many people to be seen.

Indeed, modern steel plants are highly automated.

One can look across the shed of a rolling mill without seeing a human being. But actually, machines can only work as a result of the ideas and labor of many thousand human minds and hands.

This means that some of the heaviest and most complicated tasks are not performed by people but by machines. In some parts of the mill very few workers are needed. One can look across the huge shed of a rolling mill without seeing a human being. The machines seem to be turning out bars and sheets of metal all by themselves.

Actually, though, we know that this is not really so. Machines can only work as a result of the ideas, the research, the labor, the supervision, and the maintenance of many thousand human minds and hands.

FURTHER READING

Making of Steel, The. American Iron and Steel Institute, 1000 16th Street N.W., Washington, D.C. 20036.

Meyer, Jerome S. *Iron and Steel.* Cleveland: The World Publishing Company, 1966.

Opportunities in Steel. Published by Scholastic Magazines, Inc. in cooperation with American Iron and Steel Institute. 1967.

Reference and Audio-Visual Materials on Iron and Steel: A Teacher's Guide. American Iron and Steel Institute, Washington, D.C., 6th Revision, 1973.

Tracy, Edward B. *The New World of Iron and Steel.* New York: Dodd, Mead and Company, 1971.

USA Plant Visits, 1977–1978. Published by Superintendent of Documents, U.S. Government Printing Office, Washington, D.C. 20402. (A list of nearly 2000 industrial plants around the country that welcome visitors.)

INDEX

(See Career Index on next page)

CAREER INDEX